Say What You See

Animals

Rebecca Rissman

Raintree

Chicago, Illinois

www.capstonepub.com
Visit our website to find out
more information about
Heinemann-Raintree books.

To order:

☎ Phone 800-747-4992

 Visit www.capstonepub.com
to browse our catalog and order online.

Edited by Rebecca Rissman, Daniel Nunn, and Catherine Veitch
Designed by Philippa Jenkins
Picture research by Ruth Blair
Production by Victoria Fitzgerald
Originated by Capstone Global Library
Printed and bound in the United States of America
in North Mankato, Minnesota. 052015 009003RP

Library of Congress Cataloging-in-Publication Data
Rissman, Rebecca.
Animals / Rebecca Rissman.—1st ed.
p. cm.—(Say What You See)
Includes bibliographical references and index.
ISBN 978-1-4109-5046-8 (hb)—ISBN 978-1-4109-5051-2 (pb)—
1. Animals—Juvenile literature. I. Title.
QL49.R557 2013
590—dc23   2012011705

Acknowledgments
We would like to thank the following for permission to reproduce
photographs: Shutterstock pp. title page (© Eric Isselée), 4
(© Michael Wick), 5 (© Nejron Photo, © Utekhina Anna, ©
Anatema), 6 (© Zachary Garber), 7 (© Vera Kailova, © Utekhina
Anna, © Pichugin Dmitry), 8 (© Tomas Sereda, © Christian
Musat), 9 (© Ervin Monn, © Dudarev Mikhail), 10 (© Jarek
Joepera, © FedericoPhotos), 11 (© Barry Blackburn, © Matt
Jeppson), 12 (© Krzysztof Odziomek, © Steve Noakes), 13 (©
Pichugin Dmitry, © Eric Gevaert), 14 (© intoit), 15 (© Bo Valentino),
16 (© Geanina Bechea, © Ervin Monn), 17 (© Zadiraka Evgenii),
18 (© Cheryl Ann Quigley), 19 (© Johan Swanepoel, © Wild At
Art, © Ronnie Howard), 20 (© Daniel Alvarez, © Stefanie van der
Vinden), 21 (© Ronald van der Beek), 22 (© siamionau pavel).

Cover photograph of a yellow-eyed cat reproduced with
permission of Shutterstock (© Stankevich).

Every effort has been made to contact copyright holders of
material reproduced in this book. Any omissions will be rectified
in subsequent printings if notice is given to the publisher.

Contents

Animals have a lot to say . . .
What are they saying?

Hoo, hoo

4

Woof, woof!

Howl

Grrrrrrr

5

MOO!

Hee haw!
Hee haw!

Quack, quack!

Roar!

Eee eee eee ooooh ooh!

21

Humans have a lot to say, too.

Hello, there!

Can you find these things in the book? Look back . . . and say what you see!

kitten

lion

snake

duck

Index